30119 028 772 47 5

John Lewis-Stempel is a writer and farmer. His books include the *Sunday Times* bestsellers *The Running Hare* and *The Wood*. He is the only person to have won the Wainwright Prize for Nature Writing twice, with *Meadowland* and *Where Poppies Blow*. In 2016 he was Magazine Columnist of the Year for his column in *Country Life*. He lives in Herefordshire with his wife and two children.

www.penguin.co.uk

The Wild Life of the Fox

John Lewis-Stempel

doubleday

TRANSWORLD PUBLISHERS
Penguin Random House, One Embassy Gardens,
8 Viaduct Gardens, London SW11 7BW
www.penguin.co.uk

Transworld is part of the Penguin Random House group of companies
whose addresses can be found at global.penguinrandomhouse.com

First published in Great Britain in 2020 by Doubleday
an imprint of Transworld Publishers

A CIP catalogue record for this book
is available from the British Library.

ISBN 9780857526427

Typeset in 11/14.5 pt Goudy Oldstyle Std by Jouve (UK), Milton Keynes
Printed and bound in Great Britain by Clays Ltd, Elcograf S.p.A.

Penguin Random House is committed to a sustainable
future for our business, our readers and our planet. This book
is made from Forest Stewardship Council® certified paper.

1

For my father, not least because you bought me a fox (stuffed).

The Fox

The shepherd on his journey heard when nigh
His dog among the bushes barking high;
The ploughman ran and gave a hearty shout,
He found a weary fox and beat him out.
The ploughman laughed and would have ploughed him in
But the old shepherd took him for the skin.
He lay upon the furrow stretched for dead,
The old dog lay and licked the wounds that bled,
The ploughman beat him till his ribs would crack,
And then the shepherd slung him at his back;
And when he rested, to his dog's surprise,
The old fox started from his dead disguise;
And while the dog lay panting in the sedge
He up and snapt and bolted through the hedge.

He scampered to the bushes far away;
The shepherd called the ploughman to the fray;
The ploughman wished he had a gun to shoot.
The old dog barked and followed the pursuit.
The shepherd threw his hook and tottered past;
The ploughman ran but none could go so fast;

The woodman threw his faggot from the way
And ceased to chop and wondered at the fray.
But when he saw the dog and heard the cry
He threw his hatchet – but the fox was bye.
The shepherd broke his hook and lost the skin;
He found a badger hole and bolted in.
They tried to dig, but, safe from danger's way,
He lived to chase the hounds another day.

John Clare (1793–1864)

Prologue

The Night of the Fox

I was already late when I was crossing the hall, and the phone rang. Outside, the February light was going. A conversation which should have taken one minute, wrangled into twenty.

A fatal extra nineteen minutes. By the time I put down the receiver, opened the front door to walk to the orchard, it was too late. I knew it immediately. The evening air was as dull and dead as a turned-off TV screen.

But I hoped, as you do.

When I swung open the orchard gate, I was absolutely certain. But I hoped, as you do.

As I neared the little wooden chicken shed, there were none of the usual familiar, contented murmurings of chickens settling to sleep.

Until the very last moment, when I lifted the hut door, I hoped. But all four of the chickens, Light Sussex hens, had disappeared. On the strawed floor of the hut, displayed under torchlight like an art exhibit, was the corpse of Alfreda, my daughter's pet Khaki Campbell duck. The blood on the back of Alfreda's neck was still sticky-warm to the touch.

There is always a terrible, echoing emptiness when the fox comes to visit – or, less euphemistically, slaughter. It is never believable and for a mad moment I probed the torch beam around the orchard, in case the four Henriettas – as we call our Light Sussex – had roosted in an apple tree. Up a pear tree. A cherry.

I carried on hoping, despite the tangible evidence of a duck cadaver. As you do.

My fault, of course, the death of our chickens and duck. I should have put down the phone earlier, been less concerned to satisfy some human insistence, done my job as poultry keeper. The fox was merely doing what foxes do: kill when the opportunity presents.

In that moment I despised the fox. Yet I admired her, too. The *one* time in the year I was late shutting in the fowl, she was there, killed five and took away four. To coin a phrase, I had been out-foxed.

*

Hate. Love.
Detestation. Admiration.
Livestock-killing pest. Noble hunter.

That night of the fox was only the latest of my encounters with the red fox, our largest land carnivore. I have had a lifetime of foxes. When I was a seven-year-old, a fox stole my pet bantam; the same year, I was put on the Master's horse on the Boxing Day meet at Rotherwas in Herefordshire, smelling his sloe-gin breath from the 'stirrup cup' and feeling the greasy hair of the hounds; at around ten years of age I was 'blooded' – the ritual daubing that goes back to the time of James I, where blood is smeared on the faces of those who witness their first kill when riding to hounds. (Everything about the moment was a laughable accident: I had been stuck on a pony, which required someone to walk in front opening gates, never saw the fox, and was mistaken for the intended subject of the blooding.) Clinging to the side of a tractor cab, aged twelve, the farmhand and I saw a fox foam-flecked around the mouth from the exertion of fleeing the hounds. 'Poor bugger,' said the farmhand, revving the tractor across the field and into the lane, just in time to block the hounds with the trailer full of cow muck. Even above the diesel din of the Ford County engine, we could hear the cursing of the hunt, the banging on the trailer with whip handles. The farmhand gave me a wink, and I gave him one back.

For years I had a stuffed fox in my childhood

bedroom and spent hours poring over the wonder of it – its sleekness, the precision power of its jaws, the magnificent bulk of its brush. I have watched foxes on our farm stalking snowy fields by starshine, been thrilled by their grace and beauty as they sped after rabbits in the meadow or sat and played with cubs, and once I sneaked up on a fox digging by the brook on a blowy day, and got so close that I could touch his tail and say, 'Boo!'

It was a little revenge on the fox that comes calling in the night. Despite vigilance and electric fencing, we have had poultry taken four times in the last twenty years.

As I write this, I have a print on the study wall of *Rufus Comes to Stay*. It is the only extant print of said painting by BB, aka Denys Watkins-Pitchford, the author of *Wild Lone*, probably the book that has most influenced me as a 'countryman' and writer. Aside from family mementoes, the print is my most prized possession. It depicts the winter sun rising through a stand of larch to reveal a fox on snow: in one moment, BB caught the romance and secrecy of the fox, the night hunter, who is sometimes made plain and visible by the harsh need to find food into the daylight hours.

So I have come to regard the fox with antinomies in my head, which are held together like the negative and positive terminals on a battery. *Hate/Love. Detestation/Admiration. Livestock-killing pest/Noble hunter . . .*

I find these dichotomies joltingly clarifying. Too often the fox is presented in misleading monocular in prescriptive propaganda. There are fox charities that would have one believe that the fox is without menace, where its classification 'omnivore' becomes doublespeak for 'virtually vegan'. This robs the fox of its essential brilliant truth: it is a clever killer, a dogged hunter, as efficient in dealing death as an assassin. That is why the sight of the fox, whether in a winter wood or under the orange glow of a city street light, makes the blood sing. The fox is dangerous. Wild. A being from another time: the primeval past of sabre-toothed tigers and mammoths.

Just as some people assert that the fox is all good, there are rural huntspeople and Camden Town habitués – an unlikely pairing, I grant you, but I've met them – who see the fox as 'vermin', purely as lamb-killer or polluter of patios with its excrement. Or, worse, the biter of babes in their bedrooms.

The fox is neither one thing nor the other, neither all good nor all bad. It is a red bundle of contradictions, and above all this one: it is a dog, but it refuses to be domesticated. Tauntingly and charmingly, the fox is the untameable Fido.

*

Like or loathe the fox, it has imprinted itself on British national life as much as on my own private one.

The vulpine impact is no better gauged than in place names. 'Fox' is the most popular animal-related place name in England, with 206 sites named for the species. ('Badger' comes second with 141.) Foxcombe Hill, Foxfields, Foxton . . . 'Tod' was another country word for fox, so Todmorden and Toddington, too.

Maybe you are reading this in a market-town pub. There's a good chance it's called The Fox and Hounds. Glance up at the walls. They may well be hung with prints of fox hunting. That hoard of Christmas cards you received? Even in the twenty-first century, card manufacturers consider the hunting of the fox to epitomize the traditional spirit of the British countryside.

Put aside for a moment the familiar arguments of pro- and anti-hunting, and remember the passions aroused by Labour's bill to ban the hunting of mammals with dogs. The Labour Prime Minister, Tony Blair, was never ideologically opposed to hunting. But the ban was necessary to satisfy a key block of his MPs who had helped push tuition fees through Parliament. This quid pro quo quickly became politically costly. In his memoirs he wrote, 'If I'd proposed solving the pension problem by compulsory euthanasia for every fifth pensioner I'd have got less trouble.' More than 400,000 people marched through London in September 2002 to show their support for fox hunting. What the Labour Party ended with the 2004 Hunting Act was a culture which has played a major part in British rural life for four centuries. The Thatcher years are often considered to be the great historical divide of the last hundred years. Not in the countryside. In the countryside, the great divide was the monocultural metropolitanism of New Labour, of which the Hunting Act was seized as the irrefutable proof.

Outside the light is failing, and the blackbirds of the orchard are chinking. In their den in the wood, the foxes will be waking from their sleep – it is almost their time. They belong to the night, and the night belongs to them.

So I must away to shut in the poultry and beat the fox to it. It's an old, old game.

THE CUR FOX

I

Down to Earth: The Life of the Fox

FOX. THE NOUN is from the Proto-Germanic *fuh*, meaning tail, which the fox most definitely has. Strictly, the fox which inhabits Britain is the European red fox, *Vulpes vulpes crucigera*, though it is only a lustrous burning-carmine when wearing its winter coat. Otherwise, the fox is ruddy-brown topside (hence the bookseller's term 'foxed' for the rusty staining of covers); the belly and chest are white to grey. There are black patches behind its ears and on the front of its legs. The hill fox of Scotland, an introduction from Scandinavia, is slightly larger with greyer fur. There are black foxes, tricks of genes and inheritance.

Along with the badger, *Vulpes vulpes crucigera* is perhaps our most ancient landowner. The species has lived in these isles since the geological Wolstonian period, 352,000 to 130,000 years ago.

The red fox is slender, about two feet long and fourteen inches high at the shoulder, and its long bushy tail, or 'brush', is itself about sixteen inches long and tipped with black. (The tail also inspired the word for fox in Welsh: *llwynog* comes from *llwyn*, meaning bush.) The ears are erect, the muzzle is long and pointed, with the mouth set in such a way that the fox seems to be smiling sardonically. That perpetual sneer is responsible for half the myth and misunderstanding about our wild dog.

I say dog, but one of the curiouser aspects of the fox is its strange felinity. The glinting amber eyes have vertical slit pupils, *à la* cat, and in hunting it likes to pounce like a cat. The fox walks the edge of otherness. The *kitsune*, the fox of Japanese legend, was able to shape-shift, an early capture by human art of the fox's uncertain, ambiguous nature.

I say walks, but the fox trots everywhere – except when stalking, when it may smooth itself to the ground and slink. Like a cat.

Something else about the fox's movement over earth: it treads so lightly it appears weightless. Another enigma.

The fox's den, or 'earth', is a quiet spot, under a tree root, among rocks. The fox is not a snooty, aloof neighbour. I have known foxes share a rabbit warren (with enlarged entrance) and a badger's sett. Foxes are opportunist adapters; for years an extended fox family on our farm made its den under a dystopian pile of corrugated iron (dumped in the wood by some previous agriculturalist), where they were impervious to cold and digging-out by dogs. We thought of them as the Wombling foxes, making good use of the rubbish they found. Until the 1920s, the fox in Britain was almost exclusively rural; with the spread of the Metroland semi and its long garden, complete with the handy shed under which an earth could be scratched, the fox became suburban. Having conquered the suburbs, the fox went to the town proper.

A fox's earth is untidy, with excrement and remains lying about, and a telltale musty, citrusy smell. The fox will prefer to move den, rather than clean house. In summer and fine weather, adult foxes are quite content to sleep out, especially in thorny scrub and bramble patches.

Foxes breed once a year, in January to February, when the vixen is on heat for three weeks – though she is properly fertile for three days only. A vixen screaming for a mate is the eeriest sound of the British night, together with the Pan piping of the stone-curlew. Foxes are quite vocal; aside from the

mating call, they yap and can emit a pleasingly doggy *wow-wow-wow.* The cubs are in the vixen's womb for about fifty-two days. There are usually three to six cubs in a litter (the vixen has six teats only), these born in April.

Cubs are toffee-coloured, fluffy, with blue eyes which open after two weeks. They are roly-poly playful, wall-poster cute, and since they have no innate fear of humans may be approached quite closely – until the watchful vixen shoos them to safety, deep down in their dark earth. Caution concerning humans comes later.

The dog, or male fox, will play with his offspring, as well as provide food when they are ready for solids. (The precursor to solids is food regurgitated by the vixen.) When the cubs are a month or two old, the vixen takes them hunting. These lessons in killing last until September, when the cubs are full-size. Otherwise, foxes tend to hunt alone.

In autumn the family dynamic changes, as the cubs go their separate ways to find their own territories and mates. Around 425,000 cubs are born each spring. Most do not survive their first year, and almost all foxes die before their fourth birthday. In the wild, the average life span is eighteen months.

To find a new territory, the average fox will travel the breadth of four to six territories, a distance of between nine and twenty-two miles. (In Sweden, a fox has been recorded travelling a straight-line distance of 310 miles.) The new territory will be staked out by scenting, the primary scent composed of urine laced with the compound isopentyl methyl sulphide. Truly, it is a stink, not a smell. Another scent is emitted from the violet gland near the base of the tail, so called because it has a floral aspect; foxes waft this smell around with their tail. There are also scent-producing glands in the skin between the toes and footpads, and it is the telltale secretions from these which the hounds pick up on the hunt.

Another marker of territory is excrement deposited in prominent positions, with molehills a favourite. Fox 'scat' tends to be skeleton-white, from the crunched, swallowed bones of its prey.

Size of territory can vary considerably. One of the factors determining how big a fox group's home turf will be is the availability of food. Sometimes foxes live alone, sometimes in a monogamous pair-bond, sometimes in a loose family group called a 'skulk' or 'leash', with female blood relatives helping with the raising of the cubs.

The vulpine year is strictly cyclical, an eternal circle, the beginning and end of which is the winter mating.

The fox hunts predominantly at night, though if pushed by needy young or hard weather will hunt the day long. Originally a woodland animal, the fox begins the night shift in the half-hour before dusk, when daylight is going and when the small creatures of the wood are beginning to stir and leave their lairs, and when the birds are beginning to come home to the wood to roost.

It is a good time to begin the hunt, this last

half-hour of the day, when the wood is nothing but grey shades and nothing is precise, and the fox is but a lone shadow among a thousand.

The fox can control the amount of light that passes through the lens of its eye to an extraordinary degree. In the dark the pupil is large and round; in bright light it is a perpendicular slit. Then, beneath the light-sensitive cells is the tapetum lucidum, a layer of connective tissue that allows the fox to enjoy night vision by providing twice the light that would be available normally.

The vulpine sense of smell is acute. Foxes can smell out a mammal as far as three feet below the surface. They use this excellent sense of smell alongside their ability to hear high frequencies. Humans can pick up an average of 22,000 acoustic vibrations per second (hertz) at best, but foxes can hear sounds up to 65,000 hertz. So acute is the fox's hearing – those permanently erect ears help – that it can hear a watch ticking at sixty feet.

Foxes also use their vibrissae, the sensory whiskers on their faces and legs, when out hunting. These large hairs have follicles with blood-filled sinus tissues connected to the sensory part of the brain. The fox's whiskers – proportionately longer than those of most other mammals – mean that it can 'feel' in the dark, finding its way between obstacles and along tunnels by detecting airflow.

There is yet more in the armoury of the fox: a secret weapon best seen when the red fox is out on the meadow in bleak midwinter . . .

The falling snow masks sight of the vole, but the fox can nonetheless hear the rustle through the buried grass. The fox creeps forward, with ears erect. Once it pinpoints the mouse's location, it leaps into the air to surprise its prey with a strike from above. This 'mouse pounce' has more to it than meets the human eye.

The Czech biologist Jaroslav Červený discovered that when red foxes pounce, they mostly do so in a north-easterly direction. On a north-east pounce, the fox's killing rate is 73 per cent; if they jump in the opposite direction, the success rate is 60 per cent; in all other directions, only 18 per cent of their pounces are successful. Červený suggests that a red fox may be using the Earth's magnetic field as a rangefinder, to estimate the distance to its prey and make a more accurate pounce. This targeting system works because the Earth's magnetic field tilts downwards in the northern hemisphere at an angle of 60 to 70 degrees below the horizontal. As the fox slinks forward, it seeks the spot where the angle of the sound hitting its ears matches the slope of the Earth's magnetic field. At that point, the fox knows that it is a fixed distance from its prey, and thus exactly how far to jump to land upon it.

If Červený is correct, then the red fox is unique. It would be the first animal known to use a magnetic sense to hunt, and the first to use magnetic fields to estimate distance rather than direction.

The canine teeth clamping the vole are anatomical proof of the fox's essential carnivorousness, but the fox will eat near everything it can get its spring-trap jaws on. Foxes will eat other foxes – a literal dog eat dog. A full-grown fox requires six hundred calories a day for basic maintenance. To an extent, diet is seasonal; I have seen a fox standing on its back legs eating blackberries in September, and the same dog fox up an alder tree snatching blue-tit eggs in May. (Foxes, as befits their woodland ancestry, are accomplished tree-climbers; the exceptionally long tail helps with balance.) Earthworms, crane-fly larvae and beetles are staples from spring to summer, the former sucked up from wet, warm grass like spaghetti. In winter, foxes rely mainly on smaller mammals such as mice, rabbits and, especially, voles. By preference, the fox will stalk and pounce on its prey; that said, the fox can sprint at up to thirty miles per hour and catch a running rabbit.

The sinewy strength that enables such sprinting also allows them to carry a heavy load in their jaws, such as a goose, fat for Christmas, while leaping a wire stock-fence. (As I know to my cost.) Hunting is usually done alone, though not exclusively so.

The geography and ecology of a fox's territory heavily influence its diet. Seaside foxes may subsist on crabs, fenland foxes on frogs. The red fox is one of a handful of carnivores which have adapted to living in urban centres; it exhibits what biologists term 'ecological plasticity'. While town foxes do eat wild things – those in central London probably eat more rats than their country cousins – they also scavenge from bins for leftovers. Since British society throws away about a third of the food it purchases, waste food is plentiful.

We have fed the fox, and caused a vulpine population boom in our cities. In Bristol there are thirty-seven foxes per square kilometre; in the countryside, on average, two foxes per square kilometre. Today, there are an estimated 225,000 foxes in rural areas and 33,000 in urban areas: a total of 258,000 across the land.

The country fox, of course, has also benefited from the human food-production process through its habit of eating farm livestock. This has earned it enmity in rural communities ever since humankind turned to agriculture. The earliest written account of the fox attacking livestock is from Pliny the Elder, the first-century Roman naturalist and scholar. Pliny suggested several fox deterrents for farmers. One was to put a collar of fox skin around the neck of a cockerel, which then gave the hens that mated with the cockerel immunity from fox attack. (There is perhaps more than hopeful magic at work in this remedy; foxes are territorial creatures. They rarely stray into the domain of rivals.) Nearly two millennia after Pliny, the British naturalist Brian Vesey-Fitzgerald put on the country fox's diet – aside from the inevitable chickens in the hen house – 'ducks, lambs (and full-grown sheep, if weakly), pony foals (in New Forest)'. He added that the fox 'has even been known to kill a young calf'.

Some swear blind that the gentleman in red never takes live lambs. He does. The fox is an apex predator. And apex predators take any convenient beast.

In hilly west Herefordshire, where England laps the border of Wales, we've had lambs killed by foxes. The teeth marks on the back of the neck, the nip of the spinal cord, are the forensic refutation of the fox's innocence. To be balanced on the farm books, though, is the number of rabbits a fox eats, leaving the den entrance beneath the oak ankle-deep in lapine pelts. Seven rabbits eat as much grass as one sheep.

The Burns Inquiry, a government committee set up in 1999 to examine all the issues around the debate on hunting foxes with hounds, surmised that less than 2 per cent of healthy lambs are killed by foxes in England and Wales, but remarked that 'levels of predation can be highly variable between farms and between different areas'. In the lowlands, lambs are frequently kept indoors. As many as 96 per cent of Welsh livestock farmers, according to a 2013 survey, had suffered financially as a result of foxes killing their lambs. Mountain farmers – and Wales is largely mountainous – tend to have smaller breeds of sheep, with resultant smaller lambs. The fresh-born lamb of a mountain breed is frequently no bigger than a rabbit. My own personal anecdotal evidence is that foxes take lambs in particularly pluvial times. No animal wants to hunt in endless rain.

A fox loose in the hen house or the lambing field will kill more chickens or lambs than it can cope with. Hence Vesey-Fitzgerald stating that the fox 'kills wantonly and for pure love of killing'.

To an extent, perhaps. 'Surplus killing' is common behaviour in the order *Carnivora*. A kind of blood-madness does seem to overcome the fox presented with bounty, as if the temptation is too good to resist, like a child unsupervised in a chocolate factory. However, foxes also cache and may come back for the slaughtered goods: the vulpine equivalent of filling up the larder.

Is the fox clever? Folkloric tales of the fox's cunning begin with Aesop's fables in Ancient Greece. In one fable, a crow has discovered a piece of cheese and retires to a branch to eat it. The fox flatters her by asking if her voice is as beautiful as her looks; the bird sings and drops the cheese into the fox's jaws.

According to Aelian, the Roman author writing around AD 200, foxes could sneak up to bustards by

raising their tails and pressing their chests to the ground in a persuasive silhouette of the game bird.

The *Physiologus*, a second-century Christian text, tells of the fox feigning death:

> *When he is hungry and nothing turns up for him to devour, he rolls himself in red mud so that he looks as if he were stained with blood. Then he throws himself on the ground and holds his breath, so that he positively does not seem to breathe. The birds, seeing that he is not breathing, and that he looks as if he were covered with blood with his tongue hanging out, think that he is dead and come to sit on him. Well, thus he grabs them and gobbles them up.*

(Foxes, by the way, do feign death.)

Another old tale is that the fox, troubled by fleas, takes a piece of wool in its mouth and trots to the nearest water. The fleas, desperate to avoid drowning, congregate on the piece of wool. Once they are all in place, the fox releases the wool into the water, thus getting rid of its unwelcome guests.

By the medieval era the fox's renowned cunning had entered the lexicon. In Britain the verb 'to fox', meaning to trick by craft, was in use by 1250. The Irish word for 'I play the fox', *sionnachuighim*, is where 'shenanigans' comes from. To be robbed by a fox is to be 'vulpectated'.

Shakespeare celebrated the chicanery of the fox in *Timon of Athens*: 'If thou wert the lion, the fox would beguile thee,' Timon says to Apemantus. And in *Henry IV, Part I*, Falstaff says to Mistress Quickly, in reference to the trickery of the fox, 'No more truth in thee than in a drawn fox.'

There is no time or type limit on cultural examples of the fox's guile: in BBC TV's *Blackadder Goes Forth*, Captain Blackadder asks Baldrick if his plan is 'as cunning as a fox who's just been appointed Professor of Cunning at Oxford University'.

The red fox is almost as clever as advertised. Canids have a high level of cognitive ability, and evolutionary pressure has made *Vulpes vulpes* adept at exploiting chance. The red fox, after all, is found throughout Europe, temperate Asia, northern Africa and North America, and has conquered half of Australia since its introduction there in the nineteenth century. The red fox has the largest distribution of any mammal save humans.

Everyone who has lived around foxes has a true tale to tell about the unnerving intelligence of Mr Tod. I am still flabbergasted by the fox that managed to turn off the electric fence around the chickens by pawing the wires of the tractor battery that powered it. (The fox had put its wet nose on the fence and knew the pain it caused. But how did it ascertain that the battery was the power source?) My grandfather, a fox-hunting tenant farmer, that rural stalwart now consigned to museums and books, once watched a fox escape the hounds by walking twenty yards along the top of a newly flailed hedge, putting each paw carefully on upright stubs of hazel and hawthorn. He also witnessed chased foxes double back on their tracks, enter a house, and roll in farmyard manure to fool the hounds.

Hunting foxes with dogs was made illegal in Britain from 2005. An argument for hunting was that it controlled the fox population, with about 25,000 foxes killed by hunts. Today, 80,000 foxes per year are shot by farmers and gamekeepers, and around the same number by pest-control officers in cities. Culling of foxes also takes place on conservation projects to increase or maintain biodiversity. The Loddington site of the Game & Wildlife Conservation Trust found that control of predators, including foxes, along with habitat improvement increased the density of brown hares to seventy-eight per hundred hectares; when the gamekeepers stopped controlling predators, hare density dropped to less than ten per hundred hectares. The population densities of ground-nesting birds also plummeted. The RSPB is another conservation body that sometimes culls foxes, though it prefers non-lethal deterrents. A sceptic would say that, by moving foxes on, deterrents create competition for territories, thus killing surplus foxes by the slow death of starvation.

Culling is not the main cause of death in foxes, however. A study of urban foxes in Bristol by Stephen Harris and Phil Baker reported that most fox deaths were caused by train and car contact. Disease also claimed its share, particularly sarcoptic mange, caused by the parasitic mite *Sarcoptes scabiei.* A fox with mange has patchy skin, is thin and suffers weeping eyes. The disease is contagious. In the 1990s an outbreak of mange in Bristol wiped out 95 per cent of the city's foxes. Foxes suffer other parasites and diseases, notably lungworm (*Angiostrongylus vasorum*), transmissible between individuals. Often, a fox's worst enemy is another fox.

But availability of food is the main determinant of fox life. Vixens can keep fox numbers low by suppressing their reproductive ability or practising infanticide – failing to rear live cubs.

Foxes do not make good pets. They chew and they poo.

The Fox and the Grapes

Aesop's Fables (sixth century BC)

This is the origin of the modern expression 'sour grapes'.

One hot summer's day a Fox was strolling through an orchard till he came to a bunch of Grapes just ripening on a vine which had been trained over a lofty branch. 'Just the things to quench my thirst,' quoth he. Drawing back a few paces, he took a run and a jump, and just missed the bunch. Turning round again with a One, Two, Three, he jumped up, but with no greater success. Again and again he tried after the tempting morsel, but at last had to give it up, and walked away with his nose in the air, saying: 'I am sure they are sour.'

It is easy to despise what you cannot get.

Fox-Hunting

(The Fox Meditates)

When Samson set my brush afire
To spoil the Timnites' barley,
I made my point for Leicestershire
And left Philistia early.
Through Gath and Rankesborough Gorse I fled,
And took the Coplow Road, sir!
And was a gentleman in Red
When all the Quorn wore woad, sir!

When Rome lay massed on Hadrian's Wall,
And nothing much was doing,
Her bored Centurions heard my call
O' nights when I went wooing.
They raised a pack – they ran it well
(For I was there to run 'em)
From Aesica to Carter Fell,
And down North Tyne to Hunnum.

When William landed hot for blood,
And Harold's hosts were smitten,
I lay at earth in Battle Wood
While Domesday Book was written.

Whatever harm he did to man,
I owe him pure affection;
For in his righteous reign began
The first of Game Protection.

When Charles, my namesake, lost his mask,
And Oliver dropped his'n,
I found those Northern Squires a task,
To keep 'em out of prison.
In boots as big as milking-pails,
With holsters on the pommel,
They chevied me across the Dales
Instead of fighting Cromwell.

When thrifty Walpole took the helm,
And hedging came in fashion,
The March of Progress gave my realm
Enclosure and Plantation.
'Twas then, to soothe their discontent,
I showed each pounded Master,
However fast the Commons went,
I went a little faster!

When Pigg and Jorrocks held the stage
And Steam had linked the Shires,
I broke the staid Victorian age
To posts, and rails, and wires.

Then fifty mile was none too far
To go by train to cover,
Till some dam' sutler pupped a car,
And decent sport was over!

When men grew shy of hunting stag,
For fear the Law might try 'em,
The Car put up an average bag
Of twenty dead per diem.
Then every road was made a rink
For Coroners to sit on;
And so began, in skid and stink,
The real blood-sport of Britain!

Rudyard Kipling (1865–1936)

First published in *The Strand Magazine* in 1933, Kipling's poem is modelled on the eighteenth-century song 'The Vicar of Bray'. It tells the story of a vicar who was a great survivor, living through the reigns of several sovereigns of England, with the chorus:

And this is law, I will maintain
Unto my Dying Day, Sir.
That whatsoever King may reign,
I'll be the Vicar of Bray, Sir!

Like the Vicar of Bray, the fox is a survivor. 'Fox-Hunting', a rollicking history of hunting in verse, traces the history of fox hunting from Old Testament times down to 1933, when it was written.

Living in the country as he did, Kipling was wholly aware of the mystique of fox hunting in the English countryside. In 'Below the Mill Dam' (1902), he writes approvingly of an eleventh-century abbot who 'kept the best pack in the country'. In 'My Son's Wife' (1913), he writes of a wealthy aesthete whose life is changed by inheriting a country property, where he becomes an enthusiastic rider to hounds. Nevertheless, Kipling himself did not hunt; he turned down an invitation to a day with the Pytchley Hunt, writing, 'You don't catch me outside on a hot hysterical piece of catsmeat with leather trimmings! It's vulgar.' In *An Almanac of Twelve Sports*, Kipling wrote of hunting:

Certes it is a noble sport
And men have quitted selle and swum for't.
But I am of a meeker sort
And I prefer Surtees in comfort.

('Surtees' being Robert Smith Surtees, doyen of the fox-hunt novel and creator of Jorrocks, the fox-hunting grocer.)

Still, Kipling was a stickler for accuracy; his vulpine epic, aside from using his own observations, benefited from the advice and experience of Major Guy Paget, a Northamptonshire landowner whom Kipling had met by chance at the Carlton Club. Uncannily, even in 1933 Kipling saw that the car was the great killer of the fox, rather than the pack of hounds.

II

A-Hunting We Will Go: The Fox as Sport

THE FOX HAS been on the run from humans ever since it put the wet end of its snout out of the wildwood and grabbed a lamb from the first shepherd, whose retaliation was to sling a stone. As an object for human recreation, however, the fox had to wait until the Dark Ages in Britain; King Canute classed the fox as a 'Beast of the Chase'. In 1221 Henry III gave permission to the Abbess of Barking to chase the fox in Havering Park in Essex, the earliest known fox-hunting prelate.

But throughout the Middle Ages the fox was the prey of last resort. Edward of Norwich's *The Master of Game*, written between 1406 and 1413 to instruct the future Henry V, ranked the fox seventh in the list of prey, just above the bumbling badger. Deer and hares were the prizes for the medieval hunt.

Hunting Mr Tod remained a disorganized,

minority, second-rate pastime until the Duke of Buckingham, exiled from court, regularly hunted Yorkshire's Bilsdale country with hounds in the 1670s. (In all likelihood, early fox-chasing dogs were re-educated stag-hunters; the fox-hunting cry 'tally-ho' is derived from the Norman stag-hunting *taiaut*.) The duke lived openly with the Countess of Shrewsbury after killing her husband in a duel. Fox hunting and adultery: this was two dissenting fingers to the King on each hand. The duke died of a chill caught while hunting in 1687, the first of many fox hunters to suffer a self-inflicted death through their love of the blood sport.

The fox moved to top billing for the hunting fraternity through a series of unfortunate (for the fox) incidents. The Enclosure Acts, from 1604, converted public land to privately owned land and gave England its now familiar patchwork quilt of green fields bordered by hedgerows. For hunters on horseback, Enclosure provided expanses of grassland to ride over and hedges for jumping. Also, as the deer population declined, the fox became the last animal worth hunting – really hunting – with all the paraphernalia of horses and dogs.

By the 1750s, fox hunting had become a national upper-class sport with rules, regulations, funding and, crucially, the development of faster horses and hounds.

Hugo Meynell, Master of the Quorn Hunt in Leicestershire and the 'father of fox hunting', bred foxhounds and horses for greater stamina and speed. The English thoroughbred (descended from three Eastern stallions: the Darley Arabian, the Godolphin Barb and the Byerley Turk imported in Elizabeth I's reign) went racing like the wind. Meanwhile, hounds were no longer mere anonymous dogs but Bracken and Bracket, Guilty and Gusty, Willow and Wistful. Over the grassy expanses of post-Enclosure Britain the hunt became breathtakingly exhilarating, and the hunter a charismatic figure in his red coat and top hat. The 'golden age' of fox hunting ensued, a time when the countryside was unspoilt by barbed wire and motorways. There was a surge in the number of packs of foxhounds in the second half of the eighteenth century. Only nine packs began between 1760 and 1780, but between then and the end of the century twenty-one were started.

Originally, foxhound packs were founded by the local landowning aristocracy. As the eighteenth century wore on, fox hunting spread its social net wider, to prosperous city types and to farmers. In 1792 *The Sporting Magazine* listed four packs of foxhounds within a twenty-mile radius from the centre of London. John Clare, the peasant poet from Northamptonshire, noted the effect of farmers' upward social mobility in 'The

Parish', which he wrote some time between 1822 and 1828:

> *. . . those whose clownish taste aspires*
> *To hate their farms and ape the country squires.*

Due to high agricultural prices, farmers had the money to back their social aspirations. Besides, the cost of the upkeep of their hunters could be submerged in the farm accounts.

Outside the grand shire hunts of the East Midlands, fox hunters became quite the motley crew. The novelist Anthony Trollope, who rode with the Essex Hunt from Waltham Cross, wrote in the 1860s:

> *The non-hunting world is apt to think that hunting is confined to country gentlemen, farmers and rich strangers; but anyone . . . will find that there are in the crowd attorneys, country bankers, doctors, apothecaries . . . maltsters, millers, butchers, bakers, innkeepers, auctioneers, graziers, builders . . . stockbrokers, newspaper editors, artists and sailors. Williamson explained that one of the consequences of this increased prosperity was that great landowners and local gentry began to share a single lifestyle in the countryside, mixing in a less formal, more affable way as members of a single, polite society.*

In the introduction to his long narrative poem *Reynard the Fox*, John Masefield, the Poet Laureate, was still extolling the social inclusivity of the hunt (and its kinetic beauty) in 1919:

> *But in the English country, during the autumn, winter and early spring of each year, the main sport is fox hunting, which is not like cricket or football, a game for a few and a spectacle for many, but something in which all who come may take a part, whether rich or poor, mounted or on foot. It is a sport loved and followed by*

> *both sexes, all ages and all classes. At a fox hunt, and nowhere else in England, except perhaps at a funeral, can you see the whole of the land's society brought together, focused for the observer, as the Canterbury pilgrims were for Chaucer.*
>
> *. . . Then to all Englishmen who have lived in a hunting country, hunting is in the blood, and the mind is full of it. It is the most beautiful and the most stirring sight to be seen in England. In the ports, as at Falmouth, there are ships under sail, under way, coming or going, beautiful unspeakably. In the country, especially on the great fields on the lower slopes of the Downland, the teams of the ploughmen may be seen bowing forward on a skyline, and this sight can never fail to move one by its majesty of beauty. But in neither of these sights of beauty is there the bright colour and swift excitement of the hunt, nor the thrill of the horn, and the cry of the hounds ringing into the elements of the soul.*

The development of hunting during the nineteenth century from an elite pursuit to a broader based 'subscription' pack of hounds is epitomized by the story of Jorrocks, the Cockney grocer who becomes a Master of Foxhounds (MFH) in R. S. Surtees's comic novels of the 1840s.

As hunting took off, so the fox's reputation in the

countryside was elevated. The fox was no longer vermin or lowly quarry; it was a worthy foe, hence the admiring nickname of 'Charlie' accorded the fox in the hunted countryside, after the legendarily slippery Whig politician Charles James Fox.

Ironically, hunting may have kept the species alive

in Britain, and prevented it from going the way of the wolf and the bear.

Just as change in the landscape had once teased hunting into being, hunting now brought change to the British landscape. Since Charlie needed cover to breed and feed, spinneys were encouraged and gorse coverts planted – habitats that gave both the fox and its rabbity staple meal sanctuary. Between 1800 and 1850 the amount of gorse in Leicestershire is said to have doubled. Some famous coverts which have harboured foxes for centuries are etched into rural memory, and ring through it like an incantation: Barkly Holt and John O'Gaunt's, Bilborough Gorse and Askham Bog, Rankesborough Gorse and Court House Spring.

Stimulation of fox numbers was achieved by means other than creation of these artificial coverts. There were imports of foxes from Europe; continental foxes sold at Leadenhall Market in 1845 reached ten shillings a head. Indeed, DNA analysis suggests that foxes in the south of England are more closely related to French foxes than to those from the north of England. During the eighteenth century there was a vogue for building 'fox courts', confined spaces designed to hold foxes until they were released for hunting. Sir Peter Beckford, writing in Dorset in the early 1780s, advised:

> *If you breed up [fox] cubs you will find a fox court necessary: they should be kept there until they are large enough to take care of themselves. It ought to be open at the top and walled in. I need not to tell you that it must be every way well secured, and particularly the floor of it must be bricked or paved. A few boards fitted to the corners will also be of use for shelter and to hide them. Foxes ought to be kept very clean and have plenty of fresh water: birds and rabbits are their best food.*

These foxes were rarely killed out hunting but retrieved and taken back to the kennels. Bold Dragoon was hunted thirty-six times before his eventual demise.

The fox as sporting object was not limited to being chased on horseback with hounds. In the North Country hunts were on foot, with the huntsmen wearing grey, not red – 'D'ye ken John Peel with his coat so grey?' Fox tossing, or *Fuchsprellen*, was a much-loved entertainment for aristocratic couples in seventeenth- and eighteenth-century Europe.

The sport usually took place in an arena with a circular canvas screen. Two participants would stand twenty feet away from each other while holding the ends of a cord sling. A caged fox would then be released, which ran in a mad panic around the arena. The moment it stepped on the sling, the players would pull extremely hard on the ends, hurling the animal up high into the air.

The highest recorded height for a fox toss was almost twenty-five feet.

Augustus II the Strong, King of Poland and Elector of Saxony, held a tossing contest in Dresden which used 647 foxes (plus 533 hares, 34 badgers and 21 wildcats). A further refinement in fox tossing was that the tossers would be masked, and the to-be-tossed animals likewise gaudily decorated. Classical Greece and Rome were favourite themes to dress by.

But in Britain, hunting the fox on horseback with a pack of hounds was the thing. The manners and etiquette of hunting chimed perfectly with Victorian ideals of manliness, politeness, courage. Wellington once quipped that Waterloo was won on the playing fields of Eton; he would have been more accurate to have claimed that Waterloo – and the Great War – was won on the hunting field. The British aristocracy was a warrior class, albeit in attenuated, semi-military uniform of pink (actually scarlet, but named after Thomas Pink, the eighteenth-century tailor who designed them) or black coats, white cravat and black velvet cap for the Master, Huntsman and whippers-in. The uniform of fox hunting was, and remains, precise. Identity and role are indicated by the number of buttons on the coat – five buttons for a huntsman, four buttons for a master and three buttons for a hunt member.

Irrespective of the entertainment they offered, blood sports, especially fox hunting, were means by which military skills and physical courage were taught, maintained, brushed up. Notwithstanding bone-breaking falls, the hunter was expected to get 'back in the saddle' in the manner of George Osbaldeston (1786–1866), the 'Squire of England' during the golden age of hunting who, trampled after falling and having his leg broken, carried on regardless until the end of the day. It is no coincidence that Battle of Britain pilots called 'Tally-ho!', the fox hunter's charge, when attacking the enemy.

Fox hunting also required the knack of reading landscape, and no one needs to understand landscape more than the soldier. As late as the 1930s the 5th Royal Inniskilling Dragoon Guards hunted for two months over winter.

Hunting was also bonding for country people. When Siegfried Sassoon of Weirleigh, at Matfield in Kent, enlisted in the army in 1914, he found some of his easiest hours on enlistment talking to Bob Jenner, the hunting son of a Kentish farmer:

What I should have done without him to talk to I couldn't imagine. I had known him out hunting, so there were a good many simple memories which we could share.

Before the First World War, fox hunting reached its zenith of popularity as a British field sport. But it was a very British thing, and it never travelled well. In America, the fox was allowed to escape, hence 'fox chasing'. Currently there about 180 packs of foxhounds in the UK; France has three; Portugal one.

Opposition to fox hunting developed almost synchronously with its popularity. In the mid 1700s, Whigs began to caricature the country gentry as backwoods buffoons whose sole occupation was fox hunting, and 'fox hunter' was urban slang for a West Country bumpkin. The liberal Whig caricature stuck. 'Tory equals fox hunter' is an association of notions which persists. Anti-hunting was always class war by other means, its effective slogan Oscar Wilde's quip in *A Woman of No Importance*, in which fox hunting is ridiculed as 'the unspeakable in full pursuit of the uneatable'. The (in)famous Hunt Saboteurs Association was founded in practice in 1963. Intellectually, it was born in 1763.

Less political, more philosophical, was the Victorian animal-rights movement, powered by such intellectual heavyweights as John Stuart Mill and the utilitarian philosopher Jeremy Bentham. When Bentham asked about animals, 'The question is not, Can they reason? Nor, Can they talk? But, Can they suffer?', he posed a philosophical barb that fox hunting wriggled on, and it has continued to do so to this day. In 1824 the Society for the Prevention of Cruelty to Animals was formed (later becoming the RSPCA, when given royal patronage in 1840 by Queen Victoria); badger baiting and cockfighting were both banned by 1850. Why would fox hunting be exempted from the tide of moral change?

Supporters of fox hunting relied on the old justifying card that it was 'clean-kill' pest control, citing as antique corroboration the Holderness Hunt in Yorkshire, started in 1726 by William Draper of Beswick because sheep farmers were plagued by foxes. With the widespread availability of the shotgun (which farmers are standardly allowed, precisely to kill foxes), fox hunting as a species of pest control became of next to no importance. Before the Hunting Act, registered hunting packs were estimated to kill between 21,000 and 25,000 foxes a year, a barely significant fraction of the annual 400,000-strong toll of foxes killed by gunshot, disease or road accident. As Kipling had intuited, the car was always going to be more lethal to the fox than the hound.

Fox hunting never had much to do with pest

control. The sport, which traditionally runs from 1 November to March, was always about the thrill of the chase, the music of hounds, the primacy of Country over Town. A deterrent to fox hunting, as effective as political opposition, came with the expansion of the rail and road networks, and the use of barbed wire to ring fields.

Barbed wire is dangerous to jump on a horse, especially if hidden in the hedge. The horse cannot see the barbed wire and when it attempts to brush through the top of the hedge in the usual way it will tear its belly open on the barbs. Up and up again went the cry in the hunting field of ''ware wire!' Traditional hunts lasted for hours and covered freewheeling miles; with a Britain tied in wire, it was all over by lunchtime.

Time was against fox hunting. The wonder is not that fox hunting existed, but that it lasted so long. The Burns Inquiry found that lamping with a rifle had fewer adverse welfare implications than hunting. Opinion polls showed a steady rise in favour of banning fox hunting. And so the hunting of foxes was banned in England and Wales with the Hunting Act of 2004 (Scotland had brought in a ban in 2002).

What hunts are doing now is drag or trail hunting, following an artificial scent. However, it would be

wrong to say that hunting is now a wholly bloodless sport, because foxes are still pursued to their deaths. If hounds find a fox and kill it unbidden, it is lawful. Another exemption lets hunters use a full pack of dogs to 'flush' the fox towards a bird of prey – so fox hunts now tote eagles or owls as a revanchist rejoinder.

And so fox hunting ends where it began, with the exiled Duke of Buckingham's hunt. A dissenting political act, as much as it is a sport.

The Fox at the Point of Death

A fox, in life's extreme decay,
Weak, sick, and faint, expiring lay;
All appetite had left his maw,
And age disarmed his mumbling jaw.
His numerous race around him stand
To learn their dying sire's command:
He raised his head with whining moan,
And thus was heard the feeble tone:
'Ah, sons! from evil ways depart:
My crimes lie heavy on my heart.

See, see, the murdered geese appear!
Why are those bleeding turkeys here?
Why all around this cackling train,
Who haunt my ears for chicken slain?'
The hungry foxes round them stared,
And for the promised feast prepared.
'Where, sir, is all this dainty cheer?
Nor turkey, goose, nor hen is here.
These are the phantoms of your brain,
And your sons lick their lips in vain.'

'O gluttons!' says the drooping sire,
'Restrain inordinate desire.
Your liqu'rish taste you shall deplore,
When peace of conscience is no more.
Does not the hound betray our pace,
And gins and guns destroy our race?
Thieves dread the searching eye of power,
And never feel the quiet hour.
Old age (which few of us shall know)
Now puts a period to my woe.

Would you true happiness attain,
Let honesty your passions rein;
So live in credit and esteem,
And the good name you lost, redeem.'
'The counsel's good,' a fox replies,
'Could we perform what you advise.
Think what our ancestors have done;
A line of thieves from son to son:
To us descends the long disgrace,
And infamy hath marked our race.

Though we, like harmless sheep, should feed,
Honest in thought, in word, and deed;
Whatever henroost is decreased,
We shall be thought to share the feast.

The change shall never be believed,
A lost good name is ne'er retrieved.'
'Nay, then,' replies the feeble fox,
'(But hark! I hear a hen that clocks)
Go, but be moderate in your food;
A chicken too might do me good.'

John Gay (1685–1732)

III

The Fox in Literature, Myth, Art

HUMAN CULTURE HAS been stalked by the fox ever since cavemen scrawled on walls. The Moche people of Ancient Peru conceived of the fox as a warrior who utilized brains rather than brawn. Little has changed in the ensuing millennia. The fox has been universally presented as crafty, but sometimes sinister, and frequently both.

The pillars of Western civilization are Christianity and the Classical civilizations of Greece and Rome. All promoted the villainy of the red fox. In Greek mythology the Teumessian fox was an outsize vulpine, impossible to snare. Sent by the gods as retribution, this vixen tormented the population of Thebes by eating their babes. At length, the beast was stopped when the canine Laelaps was brought in as guard dog. In the face of a paradox, where dog met and negated dog,

Zeus turned both creatures into stone and flung them into the sky. There they remain as the constellations Canis Major and Canis Minor. The Church was early to portray the fox as mendacious and ravenous; in the Old Testament Book of Ezekiel it is written, 'O Israel, your prophets are like the foxes in the desert.' Cunning, waiting only to deceive.

The fox's suborning of its brains to satisfy its gluttony became a persistent theme in Christianity. A misericord in Ely's Gothic cathedral has a carving of a fox in a preacher's gown getting close enough to the birds in his congregation to make off with one of them. A familiar, sly figure in church carvings across Europe, the fox became a byword for bestial appetite. The Tudors and Stuarts used the past-participle adjective 'foxed' to mean drunk. The fox was also depicted in literature as a rapist.

But the fox's main appearance in European myth and art comes as trickster.

In *The History of Reynard the Fox*, an epic satirical verse first recorded in Latin in the twelfth century, the trickster fox's characteristic of guile became codified and fixed on the page, and from there in the European mind for half a millennium.

There is no definite evidence as to Reynard's inventor, though in his preface to the epic poem *Reineke*

Fuchs (1794), the German writer Goethe suggests that the first account was in 1148 by Nivardus, a Flemish ecclesiast at the abbey of St Peter in Ghent. Certainly the tale of Reynard diffused throughout early medieval Europe; William Caxton derived the English translation in 1481, from the Dutch *Die Hystorie van Reynaert die Vos*, with Caxton's version in turn prompting the wicked fox in Geoffrey Chaucer's 'The Nun's Priest's Tale' of 1390. A Scottish version of *Reynard* is included in *The Morall Fabillis of Esope the Phrygian* by the fifteenth-century Scottish makar Robert Henryson, where Reynard is styled as Tod or Lowrence.

The plot of *Reynard* is straightforward. It commences in the court of a lion, King Noble, where the locals petition against Reynard's sociopathic behaviour. 'He ate my chicks,' cries Chanticleer the cockerel. 'He blinded my cubs!' exclaims Isegrim the wolf.

The King tasks Bruin the bear with bringing Reynard to court. On receiving his summons, Reynard asks Bruin if he'd like some honey. The bear sticks his head into a log to find the sweet stuff, where it gets wedged, and he loses half his head extracting it. Tybalt the cat, the court messenger, fares no better in serving the summons to Reynard. He returns with one eye.

Eventually Reynard relents and goes to the King. The remainder of the tale is a dialogue between the

two in which Reynard's skulduggeries are exposed, but, of course, he cons the King and escapes with his life.

Reynard was a phenomenon in the Middle Ages; like other medieval beast fables it was less about animals, more about human relations, using certain animals (and their natural traits) to instructively illuminate society. The tale of Reynard has since inspired homages in almost every major art form, from Stravinsky's opera-ballet *Renard* to Jánaček's *The Cunning Little Vixen*, from *The Adventures of Pinocchio* by Carlo Collodi to Ben Jonson's *Volpone*. Or Masefield's *Reynard the Fox*, one of the Poet Laureate's most popular works. In the grand tradition, Masefield's Reynard leads the hunters a merry dance and escapes in the end:

The fox lay still in the rabbit-meuse,
On the dry brown dust of the plumes of yews.
In the bottom below a brook went by,
Blue, in a patch, like a streak of sky.
There, one by one, with a clink of stone,
Came a red or dark coat on a horse half-blown.
And man to man with a gasp for breath
Said, 'Lord, what a run! I'm fagged to death.'

After an hour no riders came,
The day drew by like an ending game;
A robin sang from a pufft red breast,
The fox lay quiet and took his rest.
A wren on a tree-stump carolled clear,
Then the starlings wheeled in a sudden sheer,
The rooks came home to the twiggy hive

In the elm-tree tops which the winds do drive.
Then the noise of the rooks fell slowly still,
And the lights came out in the Clench Brook Mill;
Then a pheasant cocked, then an owl began
With the cry that curdles the blood of man.

The stars grew bright as the yews grew black,
The fox rose stiffly and stretched his back.
He flaired the air, then he padded out
To the valley below him, dark as doubt,
Winter-thin with the young green crops,
For old Cold Crendon and Hilcote Copse.

Reynard is the spivvy underdog living on his wits, the one against the many. The Caxton translation of *The History of Reynard the Fox* is written with evident glee at the character's Machiavellian quality, and his intelligence vis-à-vis the other beasts, especially those ranked above him. Says Reynard: 'An ass is an ass. Yet many have risen in the world. What a pity.'

The fox is anti-hero. In literature it tends to be domesticated dogs – Snowy in *Tin Tin*, Lassie – that get the heroic roles. An exception so rare it is worthy of mention is the 1943 novella *The Little Prince* by Antoine de Saint-Exupéry, in which the fox approaches the prince and asks to be tamed. The prince obliges, and in the process of taming the fox comes to understand the true meaning of friendship. 'He was only a fox, like a hundred thousand other foxes. But I have made him my friend, and now he is unique in all the world.'

The other role widely accorded the fox is nemesis, the eternal foe of Br'er Rabbit and Jemima Puddle-Duck.

In Britain, Reynard became immortalized in a peculiarly British art form: the fox-hunting song. 'Reynard the Fox' enjoins its listeners:

Ye gentlemen of high renown, come listen unto me
That takes delight in fox hunting by every degree
A story I will tell to you, concerning of a fox
Near royston woods and mountains high and
over stony rocks

Bold Reynard, being in his hole and hearing of
these hounds
Which made him for to prick up his ears and tread
upon the ground . . .

The famous 'A-Hunting We Will Go' was written by Thomas Arne for the 1777 production of *The Beggar's Opera* by John Gay in Covent Garden, London. From there it became a popular folk song and a nursery rhyme.

Reynard, to this day, remains a country noun for the fox.

Even more than on eighteenth-century folk song, the fox impacted on British art and novels in that century and the following one. Think of the painters who have depicted huntsmen and their horses and dogs: Stubbs, Wootton, Alken, Munnings, Herring. Virginia Woolf, no friend of the rustic, mused with puzzled admiration on the influence of fox hunters upon literature in her essay 'Jack Mytton':

> *In their slapdash, gentlemanly way, they have ridden their pens as boldly as they have ridden their horses. They have had their effect upon the language. This riding and tumbling, this being blown upon and rained upon and splashed from head to foot with mud, have worked themselves into the very texture of English prose and given it that leap and dash, that stripping of images from flying hedge and tossing tree which distinguish it.*

The most famous fox hunter in literature is the 'umble grocer Mr Jorrocks, creation of the nineteenth-century writer R. S. Surtees. Born in 1803, in 1829 Surtees was taken on as hunting correspondent by *The Sporting Magazine*, before later that year starting *The New Sporting Magazine*, which he edited until 1836 and in which John Jorrocks first appeared. The Cockney vulgarian is the unacknowledged inspiration for Dickens's Mr Pickwick.

Two years later Robert Smith Surtees's father died and he returned to the family seat at Hamsterley to live the life of a country gentleman, hunting with his own pack of hounds twice a week. He died in 1864, within ten months of his friend and illustrator Leech.

Country life in Surtees novels – the best of which are *Handley Cross*, *Mr. Sponge's Sporting Tour* and *Mr. Facey Romford's Hounds* – is inhabited by people with never a redeeming feature, a peerlessly funny gallery of parvenus and bounders. Among the truly awful is Sir Harry Scattercash of Nonsuch House, who is seldom sufficiently sober to make it to a hunt meet. If anyone believed Surtees over-inked his characters, they had never encountered the Quorn's Lord Suffield, who, on being informed that he could no longer get hay for the hunt horses on credit and that the *pâtissier* was the only tradesman who would take an order, told his stud groom, 'For God's sake, feed 'em on pastry.'

After his death, the *Dictionary of National Biography* snobbily dismissed Surtees's novels: 'Without these illustrations [Leech's] these works would have very small interest.' Surtees made the cardinal mistake for an author: he satirized the hands that would have fed him, deriding the indifference of the upper classes whilst mocking the townie nouveau riche. Neither did his bawdiness accord with Victorian prudery.

Surtees's lovingest touch with the quill was in his portraits of the English countryside – and through this vivid Edenesque landscape pads the fox, and on its tail the hunt. 'My soul's on fire and eager for the chase!' exults Jorrocks. 'By heavens, I declare I've dreamt of nothing else all night.'

Surtees gave the view from the saddle. A convincing literary account of fox life, anthropomorphism-free, came in 1938 with *Wild Lone* by BB, the pen name of Denys Watkins-Pitchford. All the better for being set near Pytchley, in the classic hunting country of Northamptonshire, *Wild Lone* privileged realism over sentiment, and gave the view of fox life from fox-eye level . . . which is nasty, brutal and short. Brian Carter's *A Black Fox Running* covered similar naturalistic territory.

Literature is a reflection of its times. Roald Dahl's *Fantastic Mr Fox* from 1970 is the story of a vulpine countercultural little man who takes on the Establishment. Mr Fox steals food from three grotesque farmers – Boggis, Bunce and Bean – through a series of underground tunnels. By organizing feasts for all the families trapped by the farmers' determination to kill the crafty fox, he is dubbed 'fantastic'.

Mr Fox is Reynard redux: the medieval trickster back, bigger and better than ever.

As a boy Dahl was anti-hunt, and wrote an essay at Repton School describing hunting as 'foolish, pointless and cruel'. *Fantastic Mr Fox* was both a mirror of the creeping change in British attitudes to fox hunting and a catalyser of the same. *Fantastic Mr Fox* influenced generations of children, just as Thomas Arne's 'A-Hunting We Will Go' once had. The 2004 ban on fox hunting owes as much to Roald Dahl of Cardiff as it does to Anthony Blair of Edinburgh.

Some Curiosities and Oddities of the Fox, Fox Life and Fox Hunting

- Red foxes have carnassials, or shearing teeth, to help cut food into manageable pieces instead of chewing.

- Foxes were introduced to Australia in the nineteenth century for recreational hunting. Today, the population of red foxes in Australia is a staggering

7.2 million. This huge number is leading to a drastic decline in the population of native mammals and birds in the country.

- The International Union for Conservation of Nature (IUCN) puts the red fox on the list of 'Least Concern' for conservation, and among the world's hundred worst invasive species.
- As well as for balance, the fox uses its bushy tail to keep warm, wrapping the brush around its head and front when lying down: its own fur coat.
- The pink or red hunting coat is the source of the idiom 'in the pink'.
- The hunt staff member known as the 'whipper-in' (of hounds) is the origin of the political terms 'whip' and 'chief whip'.
- The fox is known to produce twenty-eight different vocal sounds to communicate with its group members.
- The Inn Sign Society says that there are twenty-seven pub names containing the word fox, including The Snooty Fox, The Lazy Fox, The Crafty Fox and The Hungry Fox. (And that's before they start

counting the Tally Hos, the Horses and Hounds – and a Whipper Inn.)

- The pursuit of foxes across the countryside by hunters on horseback was instrumental in the development of steeplechasing and point-to-point racing.
- A fox cub trapped in a snare was once kept alive for two weeks by his parents bringing him food.
- Foxes are generally monogamous.
- Vulpecula, Latin for 'little fox', is a constellation first identified in the seventeenth century.
- A Tswana riddle says: '*Phokoje go tsela o dithetsenya*' – 'Only the muddy fox lives.'

IV

The Fox as Food, Medicine, Fur

Fox may well have been on the menu in Ancient Britain, though recent accounts of its consumption suggest it was too dry and gamey to be much sought-after. The fox casserole recipe of celebrity chef Clarissa Dickson Wright requires the meat to hang in running water. Then be stewed.

The various bits of the fox prescribed as medieval medicine, therefore, would have needed more than a spoonful of honey to help them down. A late Old English translation of the *Medicina de Quadrupedibus of Sextus Placitus* advises women 'who suffer troubles in their inward places, [to] work for them into a salve a foxes limbs and his grease, with old oil and with tar'. Rubbing a fox's testicles on warts was considered a means to get rid of them. Skip forward to 1607 and Edward Topsell, an Anglican priest, records many more of the fox's numerous medical uses in his *History of Four-footed Beasts*. Fox blood was prescribed for bladder stones; brains fed to infants warded off epilepsy; ashes of fox flesh incinerated and dunked in wine cured breathlessness and renal problems; fox fat was a putative cure for gout and rubbed on to bald heads as a cure for alopecia; testicles were strung around the neck of a child to cure toothache. Meanwhile, adults were encouraged to tie the penis of a fox around their aching heads to relieve migraine.

Fox fur was of mere middling importance in the skinner's trade, since it is not particularly dense (like mink), or particularly soft (like ermine). Where fox fur scored was as trimmings, because of its brilliant hue, starting in the 1500s. A century later, fox pelts exported for the European markets were fetching up to forty shillings each, and the fox-fur trade continued into the twentieth century. Fox fur reached the height of its popularity in the 1930s to 1950s, when fur stoles (complete with head and paws) and collars were popular with women, and were *de rigueur* in Hollywood movies on the shoulders of starlets. In likelihood it was the wearing of fox fur by Lana Turner, Marilyn Monroe and Grace Kelly that originated the 1940s slang 'foxy' for a sexually attractive woman.

Farms sprang up to support the new fur trade,

almost exclusively raising silver foxes, a domesticated strain of the melanistic North American red fox. Fur farming ended in the UK at the turn of the millennium, by which time there were just eleven fur farms left, most of which had long since ceased raising foxes.

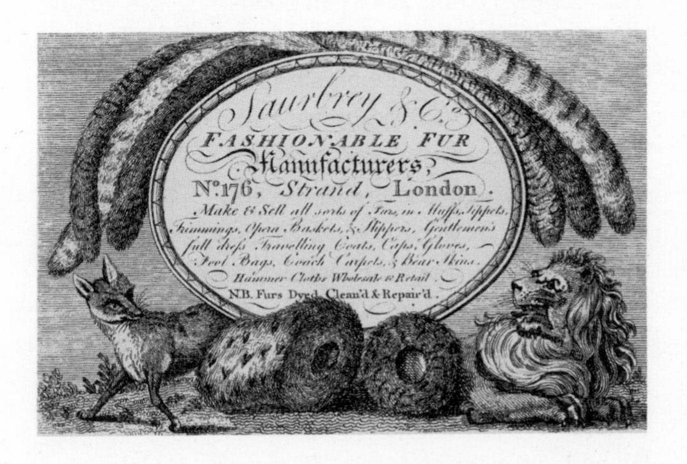

The Vixen

Among the taller wood with ivy hung,
The old fox plays and dances round her young.
She snuffs and barks if any passes by
And swings her tail and turns prepared to fly.
The horseman hurries by, she bolts to see,
And turns agen, from danger never free.
If any stands she runs among the poles
And barks and snaps and drives them in the holes.
The shepherd sees them and the boy goes by
And gets a stick and progs the hole to try.
They get all still and lie in safety sure,
And out again when everything's secure,
And start and snap at blackbirds bouncing by
To fight and catch the great white butterfly.

John Clare (1793–1864)

A Fox Reading List

BB (Denys Watkins-Pitchford), *Wild Lone: The Story of a Pytchley Fox*, 1938
E. W. Bovill, *The England of Nimrod and Surtees 1815–1854*, 1959
John Burningham, *Harquin*, 1967
Roger Burrows, *Wild Fox*, 1968
Brian Carter, *A Black Fox Running*, 1981
Raymond Carr, *English Fox Hunting: A History*, 1976
William Caxton (trans.), *The History of Reynard the Fox*, 1481
Sir Thomas Cockaine, *A Short Treatise of Hunting*, 1591
Roald Dahl, *Fantastic Mr Fox*, 1970
R. E. Egerton-Warburton, *Hunting Songs and Miscellaneous Verses*, 1859
Emma Griffin, *Blood Sport: Hunting in Britain since 1066*, 2007
Stephen Harris, *Foxes*, 1984
Stephen Harris and Phil Baker, *Urban Foxes*, 2001

Martin Hemmington, *Fox Watching: In the Shadow of the Fox*, 1997
J. David Henry, *Red Fox: The Catlike Canine*, 1986
Lucy Jones, *Foxes Unearthed: A Story of Love and Loathing in Modern Britain*, 2016
Peter Lewis, *A Fox-Hunter's Anthology*, 1935
John Lewis-Stempel, *Where Poppies Blow: The British Soldier, Nature, The Great War*, 2016
H. G. Lloyd, *The Red Fox*, 1980
Roger Longrigg, *The English Squire and his Sport*, 1977
S. P. B. Mais, *Breaking Covert: A Romance of the Hunting Field*, 1921
John Masefield, *Reynard the Fox*, 1919
David Macdonald, *Running with the Fox*, 1989
Daniel P. Mannix, *The Fox and the Hound*, 1967
Beatrix Potter, *The Tale of Mr. Tod*, 1912
Jane Ridley, *Fox Hunting*, 1990
Jonathan C. Reynolds, *Fox Control in the Countryside*, 2000
Henry Salt (ed.), *Killing for Sport: Essays by Various Writers*, 1914
Siegfried Sassoon, *Memoirs of a Fox-Hunting Man*, 1928
Roger Scruton, *On Hunting*, 1998
Thomas Smith, *Extracts from the Diary of a Huntsman*, 1921

Thomas Smith, *The Life of a Fox: Written by Himself*, 1843

E. Œ. Somerville and Martin Ross, *Some Experiences of an Irish R.M.*, 1899

R. S. Surtees, *Handley Cross*, 1843

R. S. Surtees, *Jorrocks' Jaunts & Jollities*, 1838

R. S. Surtees, *Mr. Sponge's Sporting Tour*, 1853

R. S. Surtees, *Mr. Facey Romford's Hounds*, 1865

Anthony Trollope, *Ayala's Angel*, 1881

Anthony Trollope (ill. Lionel Edwards), *Hunting Sketches*, 1865

Brian Vesey-Fitzgerald, *Town Fox, Country Fox*, 1965

A Fox Playlist

Thomas Arne, 'A-Hunting We Will Go'
Belle and Sebastian, 'The Fox in the Snow'
Big Country, 'The Red Fox'
The Jimi Hendrix Experience, 'Foxy Lady'
Bert Jansch, 'Reynardine'
Leoš Jánaček, *The Cunning Little Vixen*
Motörhead, 'Crazy Like a Fox'
Sergei Rachmaninoff, *Concerto No. 2 for Piano and Orchestra,* 'Fox Chase'
Dmitri Shostakovich, 'The Cunning Fox & The King's Daughter'
Igor Stravinsky, *Renard*
Sweet, 'Fox on the Run'
Traditional English folk song, 'The Fox Went Out on a Chilly Night'
Rick Wakeman, 'Fox by Night'

The Fox Went Out on a Chilly Night

Traditional fifteenth-century English folk song

The fox went out on a chilly night,
he prayed to the moon to give him light,
for he'd many a mile to go that night
before he reached the town-o, town-o, town-o,
he had many a mile to go that night
before he reached the town-o.

He ran till he came to a great big bin
where the ducks and the geese were put therein.
'A couple of you will grease my chin
before I leave this town-o, town-o, town-o,
a couple of you will grease my chin
before I leave this town-o.'

He grabbed the grey goose by the neck,
threw the grey goose behind his back;
he didn't mind their quack, quack, quack,
and their legs all a-dangling down-o, down-o,
down-o,
he didn't mind their quack, quack, quack,
and their legs all a-dangling down-o.

Old Mother Pitter-Patter jumped out of bed;
out of the window she cocked her head,
crying, 'John, John, the grey goose is gone
and the fox is on the town-o, town-o, town-o!'
Crying, 'John, John, the grey goose is gone
and the fox is on the town-o!'

Then John he went to the top of the hill,
blowed his horn both loud and shrill.
The fox he said, 'I'd better flee with my kill,
he'll soon be on my trail-o, trail-o, trail-o.'
The fox he said, 'I'd better flee with my kill,
He'll soon be on my trail-o.'

He ran till he came to his cosy den;
there were the little ones eight, nine, ten.
They said, 'Daddy, better go back again,
'cause it must be a mighty fine town-o, town-o,
town-o!'
They said, 'Daddy, better go back again,
'cause it must be a mighty fine town-o.'

Then the fox and his wife without any strife
cut up the goose with a fork and knife.
They never had such a supper in their life

and the little ones chewed on the bones-o, bones-o,
bones-o,
they never had such a supper in their life
and the little ones chewed on the bones-o.

Picture Credits

Alamy Stock Photo: The Natural History Museum/Alamy Stock Photo: p. ix; 19th Era/Alamy Stock Photo: p. 33; Thomas Bewick: A *General History of Quadrupeds*/Thomas Bewick: p. 8; Bridgeman Images: Private Collection/Look and Learn/Peter Jackson Collection/Bridgeman Images: p. 71; Getty Images: Whitemay/Getty Images: p. 12; DEA Picture Library/Getty Images: p. 22; The Print Collector/Getty Images: p. 27; Whitemay/Getty Images: p. 41; *Illustrated London News*, 1899/R.C. Woodville: p. 49; Shutterstock.com: Artur Balytskyi/Shutterstock.com: p. x; Anastasiia Firsova/Shutterstock.com: p. 16; Granger/Shutterstock.com: p. 58; Special Collections, University of Houston Libraries. UH Digital Library: p. 68; Wellcome Collection. CC BY: pp. ii, 34, 38; Beci Kelly: pp. 54, 76, 78, 85.

If you enjoyed *The Wild Life of the Fox*, you will love John Lewis-Stempel's previous books

MEADOWLAND

In exquisite prose, John Lewis-Stempel records the passage of the seasons in one meadowland, from cowslips in spring to the hay-cutting of summer and grazing in autumn, and includes the biographies of the animals that inhabit the grass and the soil beneath: the badger clan, the fox family, the rabbit warren, the skylark brood and the curlew pair, among others. Their births, lives and deaths are stories that thread through the book from first page to last.

THE RUNNING HARE

The Running Hare is the closely observed study of the plants and animals that live in and under plough land, from the labouring microbes to the patrolling kestrel above the corn; of field mice in nests woven to crop stems, and the hare now running for his life. It is a history of the field, which is really the story of our landscape and of us, a people for whom the plough has informed every part of life. And it is the story of a field, once moribund that becomes transformed.

THE WOOD

For four years, John Lewis-Stempel managed Cockshutt Wood, and did so in the old ways. He coppiced the trees and let cattle and pigs roam. This is his diary of the final year, by which time he knew Cockshutt from the bottom of its beech roots to the tips of its oaks, all of its animals – the fox, the pheasants, the wood mice, the tawny owl – and where the best bluebells grew. To read *The Wood* is to live among its trees as the seasons change, following the author's path. Lyrical, informative, steeped in poetry and folklore, *The Wood* inhabits the mind and touches the soul.

THE SECRET LIFE OF THE OWL

There is something about owls. They are creatures of the night, and thus of magic. But – with the sapient flatness of their faces, their big, round eyes, their paternal expressions – they are also reassuringly familiar. We see them as wise, like Athena's owl, and loyal, like Hedwig. Here, John Lewis-Stempel explores the legends and history of the owl. And in vivid, lyrical prose, he celebrates all the realities of this magnificent creature, whose natural powers are as fantastic as any myth.

THE PRIVATE LIFE OF THE HARE

The hare is a rare sight for most people. We know them only from legends and stories. They are shape-shifters, witches' familiars and symbols of fertility. They are arrogant, as in Aesop's *The Hare and the Tortoise*, and absurd, as in Lewis Carroll's Mad March Hare. In the absence of observed facts, speculation and fantasy have flourished. But real hares? What are they like? In elegant prose John Lewis-Stempel celebrates how, in an age when television cameras have revealed so much in our landscape, the hare remains as elusive and magical as ever.

THE GLORIOUS LIFE OF THE OAK

The oak is our most beloved and most familiar tree. For centuries oak touched every part of a Briton's life – from cradle to coffin. It was oak that made the 'wooden walls' of Nelson's navy, and the navy that allowed Britain to rule the world. John Lewis-Stempel explores our long relationship with this iconic tree and retells oak stories from folklore, myth and legend – oaks bearing the souls of the dead, the Green Man and fertility rites on Oak Apple Day. Of all the trees, it is the oak that speaks most clearly to us.